ASPECTS OF P.E.

THE WORKING BODY

Nuala Mullan

Heinemann
LIBRARY

First published in Great Britain by Heinemann Library
Halley Court, Jordan Hill, Oxford OX2 8EJ
a division of Reed Educational & Professional Publishing Ltd

Heinemann is a registered trademark of Reed Educational &Professional Publishing Ltd.

OXFORD FLORENCE PRAGUE MADRID ATHENS
MELBOURNE AUCKLAND KUALA LUMPUR SINGAPORE TOKYO
IBADAN NAIROBI KAMPALA JOHANNESBURG GABORONE
PORTSMOUTH NH (USA) CHICAGO MEXICO CITY SAO PAULO

Designed by Celia Floyd
Illustrations by Jeremy Gower
Printed and bound in Italy by L.E.G.O.

01 00 99 98 97
10 9 8 7 6 5 4 3 2 1

ISBN 0 431 07490 9

British Library Cataloguing in Publication Data

Mullan, Nuala
 The working body. – (Aspects of P. E.)
 1. Sports – Physiological aspects
 I. Title
 612'.044

Acknowledgements

The author and Publishers would like to thank Derek Ball, Kirk Bizley, Lynn Booth, Phil and
Pat Bradbeer, Sonia Cross, Barry Fischer, Neil Fowler, Keith George, Sue Jones, Trevor Lea,
Marie Mullan, Doug Neate and Joan Simcox for their help and comments during the
preparation of this book.

The Publishers would like to thank the following for permission to reproduce photographs:
Action Plus/Mike Hewitt p29; Action Plus/Steve Bardens p27t; Allsport/Mike Cooper p11;
Allsport/Mike Hewitt p5; Empics/Graham Chadwick p15; Empics/Michael Steele pp6, 7, 31;
Gareth Boden pp14, 21, 23; Mike Brett Photography p27b; Redferns/Mick Hutson p9; Science
Photo Library/Saturn Stills p22; Sporting Pictures (UK) Ltd p4.

Cover photograph reproduced with permission of Michael Steele/Empics

Contents

Words in **bold** in text are explained in the
glossary on page 46

The human body is made up of many complex parts. These parts must work together to form an integrated unit. As human beings, we grow, reproduce, move, obtain energy from food, react to changes and get rid of waste products. Even a simple movement such as hitting a ball with a racquet needs a complex series of events for it to occur in a co-ordinated fashion.

To understand how our bodies move, we need to study the basic structure of the body and how its parts work together so that movement can occur. This helps us to appreciate what is going on in the body when someone exercises or plays sport.

Hitting a ball requires a complex series of movements

Components of the body

Bodies come in many different sizes and shapes and each body is unique. However, the basic parts of the body are the same for everyone. The body is made up of **cells**, **tissues**, **organs** and **organ systems**:

- a cell is the basic unit of life that can maintain itself, grow and reproduce. All cells have some common characteristics. They also have their own characteristics, which make them a specific type of cell, such as a muscle or nerve cell;

- a tissue is a collection of similar types of cells. For example, muscle tissue is made up of many muscle cells;

- an organ is a collection of different tissues that work together. For example, the heart contains muscle tissue as well as connective tissue;

- an organ system is a group of organs that work together to do a certain job.

The body is made up of a number of organ systems:

- the respiratory system takes oxygen into the cells so that energy can be released;

- the circulatory system carries blood around the body, protects against disease and maintains body temperature;

- the digestive system breaks food down into small pieces so that it can enter the blood and be distributed around the body;

Many systems work together to allow movement to occur

- the nervous system senses changes, interprets them and responds. It is the main control system of the body;

- the skeletal system protects, supports and gives shape to the body. It also produces blood and allows movement to occur;

- the muscular system produces movement and helps maintain posture;

- the endocrine system releases chemicals called hormones that control various functions of the body;

- the excretory system removes waste products from the body;

- the reproductive system enables reproduction to take place;

- the lymphatic system defends the body against disease.

These systems make up the whole of the body. Together, they keep the body working properly.

Movement

To do any form of exercise we must move, and this needs a muscle contraction. For any form of movement, the muscle needs a signal telling it to contract, and energy to do this. The signal comes from the nervous system. The energy comes from food but it also needs oxygen.

The respiratory system brings air into the body and the circulatory system carries the oxygen from the air to the muscles that need it.

The digestive system breaks food down. The food is absorbed by the blood, so that the circulatory system can take it to the cells that need it.

Energy is released from the food, with the help of oxygen. The muscle can contract and the bones to which it is attached can move.

This series of events, involving the muscular system, nervous system, respiratory system, circulatory system, digestive system and skeletal system, has to take place for any movement to occur.

2 Energy for exercise

Long jump – an anaerobic event

During movement, muscles need a constant supply of energy. This energy is in the form of a substance called **adenosine triphosphate (ATP)**. There is some ATP present in the muscles, but there is not enough to supply all the energy needed for most sport or exercise. Therefore, the cells in the muscles need to make more ATP so that the muscles can contract again and again. The cells produce the necessary energy in two ways:

• **anaerobic energy** is obtained without using oxygen.

• **aerobic energy** is obtained by using oxygen.

Anaerobic energy

Anaerobic energy can be produced in two ways. The energy is provided by means of compounds called phosphates that are in the muscle cells. This energy is produced very quickly, but the stores of phosphates in the cells are small, so the amount of energy supplied is limited. This way of producing energy is called the phosphagen system and usually takes place for very short, explosive activities such as a short sprint, a javelin throw or the long jump.

The second way of producing energy without oxygen is through the **lactic acid** system. This uses carbohydrate as the fuel but does not require oxygen. Carbohydrate is stored in the muscles as glycogen. When glycogen is broken down, energy is produced, with lactic acid as a by-product.

$$glycogen \longrightarrow lactic\ acid + energy$$

Lactic acid can cause fatigue if it builds up in the muscle, so this system cannot supply energy on its own for very long. The lactic acid system is the main system used in activities where you are working at your maximum for between one and three minutes, such as a 400 m or 800 m sprint.

Aerobic energy

Aerobic energy is produced in the muscle cells, using oxygen. Glycogen is broken down to produce energy, carbon dioxide and water.

$$\begin{array}{c} glycogen \\ + \\ oxygen \end{array} \longrightarrow \begin{array}{c} carbon\ dioxide \\ + \\ water \\ + \\ energy \end{array}$$

Fat can also be broken down in the presence of oxygen to produce energy. The aerobic system is a lot slower than the anaerobic system, but it gives you a plentiful supply of energy. The carbon dioxide produced is removed from your body through the respiratory system. The aerobic system is the main energy system for longer, less intense endurance activities, such as the marathon.

Oxygen uptake

The amount of oxygen that your body takes in and uses at any time is called the volume of **oxygen uptake (VO²)**. Your oxygen uptake increases as you work harder. Everyone has a maximum ability to take in and use oxygen. This is called their maximum oxygen uptake or VO_2 max. Endurance training will improve your maximum oxygen uptake, but some people are able to achieve a higher VO_2 max than others. This is due to physical differences that are inherited from your parents. The top endurance athletes have very high VO_2 max values as a result of hard training but also because they inherited the right physical characteristics from their parents.

The three energy systems described before work together to supply energy. Which system your body brings into play will depend on how long and how hard you exercise.

Recovery process

After exercising, your breathing rate stays high (more oxygen is breathed in) and your heart continues to beat fast. Gradually both of these will return to normal. The term **oxygen debt** is used to describe the need for extra oxygen uptake (above resting levels) that occurs during recovery.

During recovery from exercise, the breathing rate remains high

Your body uses extra oxygen to release energy to rebuild the anaerobic compounds that were broken down at the start of exercise and to help remove lactic acid from your muscles.

Fact File

Cool-down: a cool-down (also called a warm-down) aims to bring the heart rate and oxygen uptake gradually back to resting levels. During a cool-down you continue to exercise at a gentle pace that keeps the blood circulating around the body faster than if you stopped suddenly. The circulating blood removes the unwanted lactic acid from your muscles more quickly and helps reduce muscle stiffness after exercise.

3 The respiratory system

To keep working, the body needs to get air inside and supply oxygen to wherever it is needed. The body also needs to get rid of some gases that build up inside. Together these are the jobs of the respiratory system. The respiratory system is made up of several organs.

Mouth and nose

Air enters your body through your mouth (**buccal cavity**) or nose (**nasal cavity**). These are separated by the **palate** at the top of the mouth. These two cavities join up at the back of your mouth and air passes from both of them into the **pharynx** (throat).

When air enters your nasal cavity through the nose, it is warmed, moistened and filtered. Hairs inside the nose trap any large dust particles. The air then passes into the upper nasal cavity that is lined by **mucous membranes**. These membranes release a thick fluid called **mucus**. The mucus moistens the air and traps smaller dust particles, preventing them from travelling further into your body.

When you are resting, you breathe most air in through your nose. During exercise you need to get a lot of air inside, so you breathe air in through both your mouth and nose.

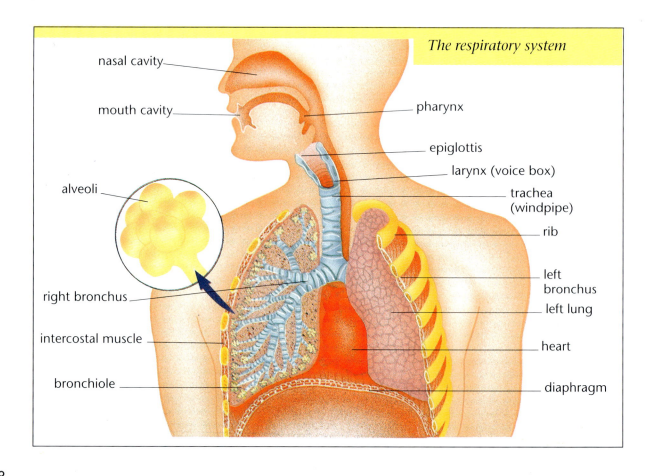

The respiratory system

nasal cavity

mouth cavity

alveoli

right bronchus

intercostal muscle

bronchiole

pharynx

epiglottis

larynx (voice box)

trachea (windpipe)

rib

left bronchus

left lung

heart

diaphragm

Pharynx

Your pharynx, or throat, is a passageway for air and food. It is also where you produce speech sounds. Food passes towards the food pipe (**oesophagus**) and air passes towards the windpipe (**trachea**) via the voice box (**larynx**).

Larynx

Your larynx is a short passageway connecting your throat to your windpipe. Inside the larynx are the vocal cords. These are folds of mucous membrane that vibrate when air passes through, producing sound waves. The sound then resonates in your pharynx, nose and mouth and is converted into speech.

Epiglottis

At the top of the larynx is a flap called the **epiglottis** which acts like a gate. Its job is to allow air to pass into the larynx and to prevent food from getting through. When you swallow, the epiglottis closes over the top of the larynx so that food goes down the food pipe and not into the larynx. If your food does go down the wrong way, the process has not worked and you cough to bring the food back up into the throat.

Trachea

Your trachea, or windpipe, is a passageway (12 cm long) which connects your larynx to your lungs. It is surrounded by incomplete rings of cartilage that prevent the wall from collapsing in and obstructing airflow. When you swallow, the food pipe expands slightly into the gaps in the cartilage of the trachea.

The vocal cords are used to produce the sounds in singing

Fact File

The pitch of the voice is controlled by the tension in the vocal cords. When the tension is greater, the pitch is higher. The vocal cords in men are usually thicker and longer and vibrate more slowly than in women. This is why men's voices generally have a lower pitch than women's.

Bronchi and bronchioles

The trachea divides into two passageways that lead into the lungs. The right **bronchus** leads to the right lung and the left bronchus leads to the left lung. Together they are called **bronchi**. Because it is shorter, more vertical and wider than the left bronchus, foreign objects are more likely to enter the right bronchus. These bronchi divide into smaller bronchi in the lungs, and these divide into smaller tubes called **bronchioles**. This network of tubes is called the bronchial tree because the network of passageways looks like the branches of a tree.

Alveoli

The bronchioles lead into millions of tiny air sacs called **alveoli**. Gas exchange of oxygen and carbon dioxide takes place in the alveoli.

Lungs

Your bronchi, bronchioles and alveoli together make up your lungs. The two lungs are the link between the outside air and the blood. The liver lies below the right lung, which is shorter than the left lung.

Each lung is surrounded by a double-layered membrane called the **pleural membrane**. The outer layer is attached to the wall of the rib cage (**thoracic cage**) and the inner layer covers the lungs. The space between these layers is called the **pleural cavity**. The membranes release a lubricating fluid into the pleural cavity to allow them to move smoothly against one another during breathing. If these layers become inflamed, it causes friction between them and can cause pain.

Thoracic cage

The lungs are surrounded by your thoracic cage, which protects the lungs and heart from physical damage. The thoracic cage is made up of twelve pairs of ribs. In between the ribs are two layers of muscles called the **internal and external intercostals**, which are involved in the breathing process. The space inside the thoracic cage is called the thoracic cavity.

Diaphragm

Beneath your lungs is a large, dome-shaped muscle called the **diaphragm**. It separates the thoracic cavity from other organs such as the stomach and liver. This muscle plays an important role in the breathing process.

Fact File

Coughing is a reflex action that sends air rushing through the respiratory passages. A cough can be triggered by a foreign body lodged in the trachea, larynx or epiglottis.

A sneeze is set off by an irritation of the mucus in the nose. It is a sudden contraction of the internal intercostal muscles and it forcefully expels air through the mouth and nose.

A hiccup is a sudden contraction of the diaphragm and rapid closure of the epiglottis, producing a sharp sound. It is usually triggered by irritation of the nerves associated with the digestive system.

These waterpolo players are breathing hard to get air into their lungs

The breathing process

The breathing process brings fresh air into your lungs and removes waste products such as carbon dioxide and water vapour from your body. Breathing in is called inspiration (or inhalation) and breathing out is called expiration (or exhalation).

The movement of air into and out of your lungs is called ventilation. When resting, the breathing rate of an adult is usually twelve breaths per minute. During exercise this rate increases to get more air into the lungs.

Inspiration

When you breathe in, this is what happens:

- the diaphragm muscle contracts and flattens;
- the external intercostal muscles contract and pull the rib cage upwards and outwards;
- the volume inside the ribs increases and the pressure decreases;
- air flows into the lungs.

Expiration

When you have finished breathing in, this is what happens:

- the diaphragm and intercostal muscles relax;
- the dome of the diaphragm moves upward and the rib cage moves downward and inward;
- the volume inside the ribs decreases and the pressure increases;
- air flows out of the lungs.

Normally, breathing out is a passive process. That means it is not caused by muscles contracting. However, when you exercise hard, expiration becomes active. You need to use your stomach muscles and internal intercostal muscles to help force air out of your lungs.

Fact File

The Heimlich Manoeuvre is a first-aid technique used on someone who is choking. The first-aider stands behind the casualty and places both fists against their upper stomach. The first-aider then quickly thrusts both fists up against the abdomen. This causes a rapid increase in pressure in the lungs and forces air rapidly through the trachea and should expel the foreign object. However, this technique can be dangerous so should only be carried out by a qualified first-aid person.

Composition of air

The air that we breathe is made up of a mixture of gases including nitrogen, oxygen and carbon dioxide. Inspired air is made up of 79.04 per cent nitrogen, 20.93 per cent oxygen and 0.03 per cent carbon dioxide.

When you take a breath in, these gases enter the mouth and nose and travel along the network of passageways into the alveoli. Here the new air mixes with the air that is already in the alveoli, from the last breath. This changes the composition of the air in the alveoli. Oxygen and carbon dioxide gas are exchanged, between air and blood. This alters the levels of these gases in the air in the alveoli.

When you breathe out the composition of expired air is 79.61 per cent nitrogen, 16.89 per cent oxygen and 3.5 per cent carbon dioxide. This shows that the lungs are not very efficient at extracting oxygen.

For someone at rest, expired air still contains almost 17 per cent oxygen. During exercise, though, the amounts of oxygen taken into the body and carbon dioxide released will increase.

Gas exchange

Inside the lungs, the alveoli are surrounded by millions of tiny blood vessels called **capillaries**. During gas exchange, oxygen passes from the alveoli to the blood in the capillaries and carbon dioxide passes from the blood to the alveoli. This movement of particles from an area of high concentration to an area of low concentration is called **diffusion**.

The air in the alveoli is rich in oxygen and low in carbon dioxide, whereas the blood in the capillaries is low in oxygen and rich in carbon dioxide. As a result, oxygen diffuses from the air in the alveoli to the blood in the capillaries, and carbon dioxide diffuses in the opposite direction. Oxygen is then carried around your body to where it is needed and carbon dioxide is exhaled from the lungs.

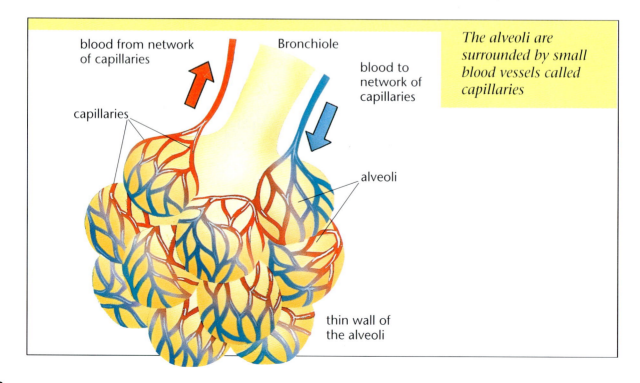

blood from network of capillaries

Bronchiole

blood to network of capillaries

capillaries

alveoli

thin wall of the alveoli

The alveoli are surrounded by small blood vessels called capillaries

The alveoli have several features which help gas exchange:

- *large surface area* – if all the alveoli in a pair of lungs were opened up and laid out flat on the ground, they would take up an area half the size of a tennis court! This means that there is a large surface for diffusion between the alveoli and blood;

- *thin walls* – so diffusion of gases through the walls is easier;

- *moist lining* – helps with the diffusion of gases;

- *surrounded by many capillaries* – also with thin walls.

Effects of exercise

During exercise your body needs more oxygen to increase the energy production in the working muscles. To achieve this, you need to increase the rate and depth of breathing. This increases the amount of air moving into and out of your lungs (ventilation) and the amount of oxygen that is taken into your blood. As a result of the greater energy production, more carbon dioxide is produced and exhaled from your lungs.

There are two main, long-term benefits of exercise on the respiratory system. After training, the maximum amount of air that you can breathe in and out increases. This is called increased maximum ventilation, and it allows you to get more oxygen to the muscles and to do more exercise. Secondly, after training your body is more efficient at taking oxygen in and using it, so you do not need to breathe as much at a particular activity level than you did before. Both of these changes result from your improved aerobic fitness.

Smoking

Although smoking cigarettes has been proven to be very harmful to health, it is still a common habit among many adults and young people. Cigarette smoking

- reduces resistance of the lungs to disease and often causes cancer;

- irritates the passageways of the lungs and gradually destroys their lining;

- increases airway resistance making breathing harder;

- can be harmful to the baby of a pregnant mother.

Smoking also increases the demand for oxygen when exercising, because extra energy is needed for breathing. For anyone who does smoke, this effect can be reduced by not smoking for a day before a big competition.

Fact File

Expired air resuscitation (EAR) is used to revive a casualty who has stopped breathing. The first-aider breathes expired air into either the mouth or nose of a person who has stopped breathing. The oxygen level in expired air (17 per cent) helps keep the casualty alive and the carbon dioxide level (3.5 per cent) stimulates the casualty to start breathing again. Only a qualified first-aider should try this, though.

Measuring lung volumes

Several techniques can be used to measure different types of lung volume, and to examine how well the lungs are functioning.

Total lung capacity

This is the total amount of air that can be held in your lungs. It depends on the size of your lungs. The total lung capacity is made up of the **vital capacity** and the **residual volume**.

Vital capacity

This is the largest amount of air that can be expired after a maximum inspiration. You can measure your vital capacity by taking a deep breath in and then blowing as much air as possible into a **spirometer**, a device for measuring gas volumes. The average value for the vital capacity of adults is 3–4 litres for females and 4–5 litres for males.

Residual volume

Even when you breathe out as hard as you can, not all the air is expelled from your lungs. The residual volume is the volume of air that is left in the lungs after a maximal expiration. This is not easy to measure because the air remains in the lungs. The residual volume is normally about one-quarter of the vital capacity. Average values for residual volumes of adults is 1–1.2 litres for females and 1.2–1.4 litres for males.

Tidal volume

This is the amount of air that you breathe in or out in one breath. At rest the tidal volume of an adult is about 0.5–1 litre, which means that 0.5–1 litre of air is breathed in or out in each breath. During exercise this can increase dramatically with the need to get more air into the body. You can achieve this by using some of the reserve volumes:

- **inspiratory reserve volume** – the amount of air that can be forcefully inspired at the end of a normal inspiration;

- **expiratory reserve volume** – the amount of air that can be forcefully expired at the end of a normal expiration.

It is possible to measure these volumes but they are not a good indicator of fitness. The size of the lungs cannot be altered by fitness training. The strength of the respiratory muscles can be affected, especially in sports such as swimming and scuba diving, where breathing is more forced and so the respiratory muscles work harder. However, the main influence on fitness is how efficient your respiratory system is at taking in oxygen and expelling carbon dioxide.

Measuring lung volumes can, however, give an indication of the health of the lungs and surrounding structures. Abnormal function may be the result of smoking or asthma. One test that is commonly used with asthma sufferers is the peak expiratory flow test.

Peak expiratory flow

This is a measure of the maximum flow rate at which air can be expired from the lungs. It can be measured by blowing as fast as possible into a peak flow meter, which measures the speed of the air. If the flow of air is restricted in any way the value for peak flow will be reduced.

Asthma

Asthma is a respiratory disease characterized by shortness of breath, wheezing and coughing. The airways in the lungs become narrow and this makes breathing difficult, a bit like breathing through a thin tube. People with asthma are extra-sensitive to certain irritants such as pollen, dust and cold air and to exercise. Most of the time, people suffering from asthma can breathe normally, but when an irritant affects the lungs it can bring on an asthma attack. This makes it difficult for the person to breathe out, so it is difficult to get air into the lungs. When this happens the person can start to breathe very rapidly and gasp for breath.

Most asthma sufferers use an inhaler. This is a simple device that allows the asthmatic to inhale a dose of medicine that will cause the airways to enlarge, allowing breathing to return to normal. During exercise, more air has to be breathed in and out and this can trigger an attack, especially in cold weather. That is why you may see a person with asthma using their inhalers before PE lessons.

This helps reduce the chances of an attack during the lesson. Many asthmatics use a peak expiratory flow meter daily to monitor the state of their airways. A lower-than-usual reading indicates that the airways are narrowed. The amount of medication taken can then be adjusted accordingly.

Ian Botham has asthma but still became a top-class cricketer

4 The circulatory system

The circulatory system is made up of the heart, blood and blood vessels. Its three main functions are:

- *transport* of nutrients (food) around the body and of oxygen and carbon dioxide to and from the lungs;

- *protection* against disease by the production of antibodies;

- *temperature regulation* by transporting heat from the centre of the body (core) to the skin or from the skin to the core. This is done to keep body temperature stable.

The network of blood vessels in the circulatory system ensures that blood reaches all parts of the body, carrying oxygen to working cells and carbon dioxide away from them.

Heart

Your heart is the powerhouse of the circulatory system. Blood passes through the heart on its way to the tissues and again on its way back from the tissues, then onto the lungs. Each time your heart contracts, blood is pumped out and sent on its way.

The heart is made up of a specialized type of muscle called cardiac muscle. The structure of the heart consists of:

- **atria** (or auricle) – the two upper chambers, right atrium and left atrium;

- **ventricles** – the two lower chambers, right and left ventricle;

- **bicuspid (mitral) valve** – a one-way valve separating the left atrium from the left ventricle;

- **tricuspid valve** – a one-way valve separating the right atrium and right ventricle.

The valves ensure blood flows through your heart in the correct direction. The right and left sides of the heart are separated by a wall so that blood cannot pass between one side and the other. Because of this the heart acts like two separate pumps.

Blood flow through the heart

Blood that has come from the lungs is rich in oxygen and is called oxygenated blood (coloured red in the diagram). This oxygenated blood follows a circuit around the body:

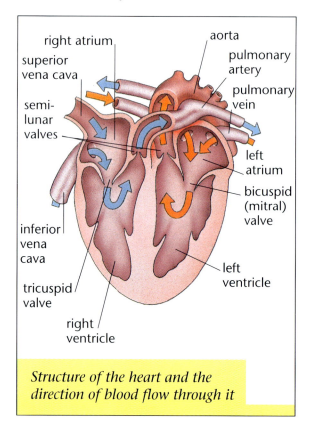

Structure of the heart and the direction of blood flow through it

- oxygenated blood enters the left atrium of the heart via large blood vessels called the **pulmonary veins**.

- the left atrium contracts and pushes the blood through the bicuspid valve into the left ventricle.

The left ventricle then contracts and forces the blood out of the heart through a **semilunar valve**. The blood leaves the heart via a large blood vessel called the **aorta** that distributes the blood around the body;

- the tissues around the body take oxygen out of the blood. This blood is now called deoxygenated blood (coloured blue in the diagram);

- deoxygenated blood returns to the heart via two large blood vessels, called the **superior and inferior vena cava**, which empty into the right atrium;

- the right atrium contracts, forcing blood into the right ventricle through the tricuspid valve;

- the right ventricle contracts, pumping blood out of the heart through a semilunar valve. The deoxygenated blood leaves the heart via the **pulmonary artery** and returns to the lungs. Here carbon dioxide is exchanged for oxygen and the circuit is repeated.

Because it is constantly circulating around the body, blood is passing through both sides of the heart at the same time. Both atria contract at the same time, then both ventricles contract at the same time. Together these events make up a heart beat. The heart is then relaxed until the cycle begins again for the next heart beat. During exercise this circuit has to be completed much more quickly, so the period of relaxation between beats is much shorter.

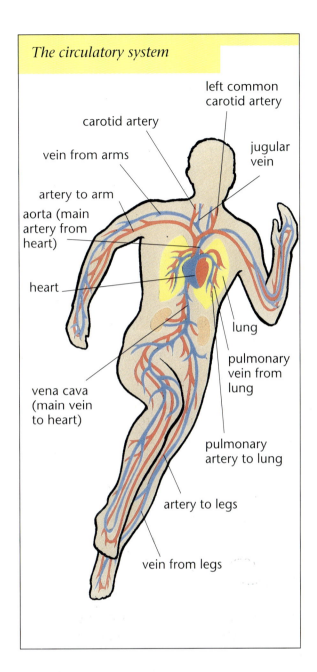

The circulatory system

left common carotid artery

carotid artery

vein from arms

jugular vein

artery to arm

aorta (main artery from heart)

heart

lung

pulmonary vein from lung

vena cava (main vein to heart)

pulmonary artery to lung

artery to legs

vein from legs

Fact File

A normal resting heart rate for an adult is around 70 beats per minute. This means that the heart beats over 100,000 times per day, pumping 7000 litres of blood around 96,000 kilometres of blood vessels.

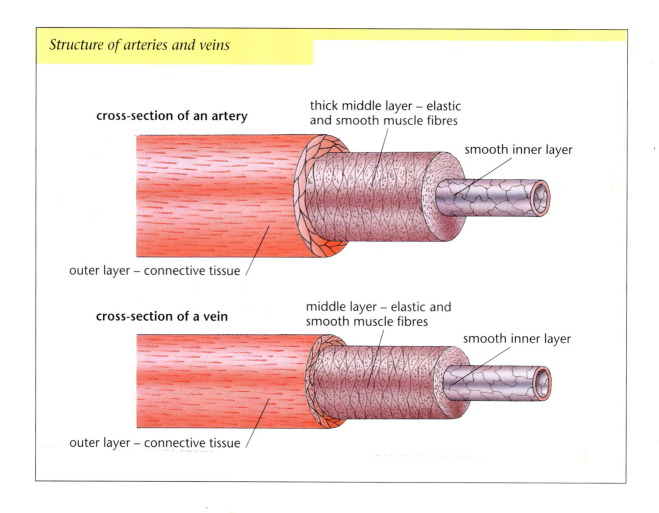

Structure of arteries and veins

cross-section of an artery

thick middle layer – elastic and smooth muscle fibres

smooth inner layer

outer layer – connective tissue

cross-section of a vein

middle layer – elastic and smooth muscle fibres

smooth inner layer

outer layer – connective tissue

Pulmonary and systemic circulation

The flow of blood around your body is called the circulation:

- the **pulmonary circulation** is the circuit of blood between your lungs and heart. It carries deoxygenated blood from the right ventricle to the lungs where carbon dioxide is released and oxygen taken up from the alveoli. The oxygenated blood is then carried from the lungs to the left atrium of the heart;

- the **systemic circulation** is the link between your heart and the rest of your body. Oxygenated blood leaves the left ventricle and is transported around

your body to your muscles, brain, kidneys, liver and other tissues. Your organs and tissues extract oxygen from the blood and deposit carbon dioxide. The deoxygenated blood then returns to the right atrium of your heart.

The walls of the left ventricle have a thicker layer of muscle than the right ventricle because it has to pump blood around your body whereas the right ventricle only has to pump blood as far as your lungs.

Blood vessels

Blood vessels are a network of tubes that carry blood around your body. There are different types with individual functions.

Arteries

Arteries are blood vessels that carry blood away from your heart. The pulmonary artery carries blood from the heart to the lungs. The **aorta** is an artery that carries blood from the heart to the rest of your body. This large artery divides into smaller arteries that branch out in different directions. These then subdivide into even smaller arteries called **arterioles**, which lead to tissues throughout the body. The centre of an artery or **vein** that the blood flows through is called the **lumen**.

The walls of arteries have three layers:

- a smooth inner layer;
- a thick middle layer of elastic and smooth muscle fibres;
- an outer layer of connective tissue.

The thick muscle layer means that the arteries can cope with the pressure of the blood being pumped out of your heart. It also means that blood flow along an artery can be controlled. When the muscles contract the lumen is reduced and less blood can flow through. This is called **vasoconstriction**. When the muscle layer is relaxed, the size of the lumen is increased and more blood can flow through. This is called **vasodilation**. These processes allow the amount of blood flowing to different parts of your body to be controlled.

Veins

Veins are blood vessels that carry blood to your heart. Small veins called **venules** join up to form larger veins. The pulmonary vein carries blood to the heart from the lungs. The superior and inferior vena cava carry blood to your heart from around your body.

The walls of veins have the same three layers as arteries but the thickness of the layers is different. The middle and outer layers are thinner than in an artery.

One major difference between veins and arteries is that veins have valves. Blood that is returning to the heart from around your body is at a lower pressure. This is not enough to push the blood back to the heart. The valves in your veins prevent any backflow of blood. In the upper part of your body, gravity helps push blood back to the heart. The contraction of the muscles surrounding the veins also helps squeeze blood back to your heart.

Capillaries

Capillaries are the connection between arterioles and venules. They form a network of very fine tubes around tissues and allow small particles to pass from the blood into the tissues and vice versa. The walls of capillaries are only one cell thick, which makes it easier for **diffusion** to take place.

Fact File

In some people, fat may accumulate on the walls of the arteries and partially obstruct blood flow. The medical name for this build-up of fat is atherosclerosis. People who smoke, eat a lot of fatty food or are inactive are at greater risk from atherosclerosis. When this happens in the arteries of the heart, it is called coronary artery disease or heart disease.

Blood

Blood plays a vital role in transporting substances such as oxygen around your body, regulating temperature and fighting disease. On average an adult male has 5–6 litres and an adult female has 4–5 litres of blood in their body. For a person with a blood volume of 5 litres, at any one time 1 litre will be in the lungs, 3 litres in the veins of the systemic circulation and 1 litre in the heart, systemic arteries and capillaries. Blood is mainly made up of:

● red blood cells

● white blood cells

● plasma.

Red blood cells

Red blood cells (**erythrocytes**) normally make up 40–47 per cent of blood. Their main function is to transport oxygen. **Haemoglobin** is a component of red blood cells and when it combines with oxygen, **oxyhaemoglobin** is formed. This is how oxygen is transported in the blood. Oxyhaemoglobin is a bright red compound that gives blood its colour. Red blood cells also play an important role in the transport of carbon dioxide.

Having more red blood cells means that more oxygen can be carried in the blood, which helps increase aerobic fitness. Some athletes train at high altitudes, which helps them develop the ability to carry more oxygen in their blood. However, the change is not permanent. Blood doping is a procedure which increases the number of red blood cells. Blood is removed from the athlete and stored for a period of time. During this time the athlete's body makes more red blood cells to replace those lost. The stored blood is then injected back into the athlete, to raise the number of red cells in the blood. It is illegal for athletes to use this procedure.

White blood cells

White blood cells (**leucocytes**) are outnumbered by red blood cells in a ratio of 600 to 1. They defend your body against disease by attacking foreign bodies or producing antibodies, substances that attack and destroy the foreign bodies.

Platelets

The third type of cells in the blood are called platelets (**thrombocytes**). They play an important role in blood clotting.

Table of characteristics of blood cells

Cells	Number per mm³	Produced in	Life span
Red (erythrocytes)	5 million	Bone marrow	120 days
White (leucocytes)	5–10,000	Bone marrow and lymphatic tissue	Few hours to few days
Platelets (thrombocytes)	250–400,000	Bone marrow	5–9 days

Plasma

Water is the main component of plasma (91 per cent) along with plasma proteins (7 per cent), salts and other substances. The plasma proteins have various functions including production of antibodies to fight disease. They also aid in the process of blood clotting.

Taking the radial pulse

Blood-clotting mechanism

Platelets gather and break down at the site of an injury. They release a chemical that results in the production of a substance called fibrin. Fibrin forms a network of long strands that trap blood cells and form a clot. This stops blood from escaping from a cut in your skin.

Pulse locations

When your heart beats, blood is pumped around your body and causes the walls of the blood vessels to expand and then relax. This expansion is your pulse and can be felt at the following sites on your body:

- **carotid pulse** – just below the jaw on either side of your neck near the throat;

- **femoral pulse** – in the groin area on either side of your pubic bone;

- **radial pulse** – on the inside of your wrist just below the base of the thumb;

- **temporal pulse** – on either side of your forehead in the temple area.

Fact File

Haemophilia is a 'bleeding disease'. This means that when a person's skin is cut, a clot does not form and bleeding continues. It can result in major blood loss and serious problems. People with haemophilia do not have a special substance that plays an important part in the breakdown of platelets. Without this substance the platelets do not break down and clotting does not occur.

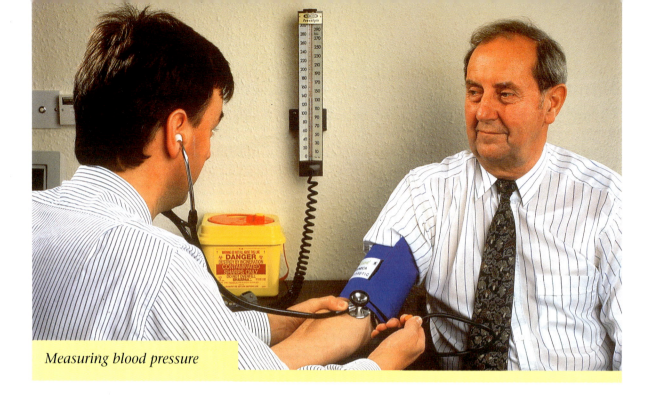

Measuring blood pressure

Blood pressure

When your heart contracts it forces blood out of itself and into the blood vessels. The force that the blood exerts on the walls of your blood vessels is called the **blood pressure**. Blood pressure changes depending on how much blood is pumped out of your heart and the resistance that the blood meets travelling through the blood vessels. As the blood travels away from your heart, friction between the blood and the walls of the arteries causes the blood pressure gradually to fall. Therefore, the blood pressure in the capillaries and veins is much lower than in the arteries.

In a healthy young person, the arteries are elastic and blood flows easily through them. As people get older this elasticity is lost and the resistance to blood flow is greater. This means that the heart has to pump harder to get blood around the body, so the blood pressure increases.

Blood pressure can be measured by using a stethoscope and a **sphygmomanometer**. This device

consists of a cuff, an inflation bulb and a mercury column for the pressure readings. The cuff is placed around the person's arm and inflated until the blood supply is cut off. The pressure in the cuff is then slowly released and two pressure readings are taken when certain sounds are heard through the stethoscope. The two readings are:

- **systolic pressure** – the pressure when the heart is contracting;

- **diastolic pressure** – the pressure when the heart is relaxing.

The blood pressure reading is given as the systolic reading over the diastolic reading. An average blood pressure reading is around 120 over 80 (120/80). The units are millimetres of mercury or mmHg.

Effects of exercise

When you exercise, your working muscles need a greater supply of oxygen and, therefore, blood. A number of changes occur in the circulatory system to achieve this.

Heart rate changes

Your heart rate increases during exercise, depending on the intensity of the activity. An increase in exercise intensity causes a directly proportional increase in heart rate. However, each person has a maximum heart rate. When this is reached, the heart rate cannot increase further even if the exercise becomes harder. Exhaustion will occur soon after maximum heart rate is reached. Your approximate maximum heart rate can be calculated as follows:

maximum heart rate = 220 – age

For example, if you are 15, your maximum heart rate is 205.

Heart rate levels are often used by athletes to set their training intensity.

Blood pressure

During exercise the amount of blood pumped out of your heart per minute is increased. This means that the heart is pumping harder as well as faster and so blood pressure increases. High blood pressure when you are resting is a sign of health problems, but you need a high blood pressure during exercise, to help circulate the blood. This is a perfectly healthy situation.

Blood flow

At rest, all areas of your body receive enough blood flow to ensure they have sufficient oxygen to work. During exercise the working muscles need a much greater blood supply. This is achieved in two ways. Firstly, the heart rate increases and blood is circulated around your body at a faster rate, increasing blood flow. Secondly, the distribution of blood around the body is changed. The blood vessels leading to the

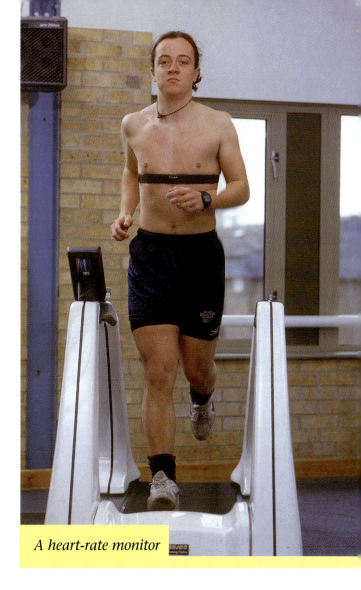

A heart-rate monitor

working muscles expand (vasodilate) so that the extra blood flow goes to these muscles. The blood vessels leading to other areas contract (vasoconstrict) so that only the necessary amount of blood flows to these parts. This ensures that the extra blood goes to where it is needed the most.

Temperature

The amount of heat generated in your body increases during exercise. Your body has to get rid of this excess heat, to keep your body temperature stable. The heat is transported to your skin by your blood and is dispersed. During exercise or in hot weather, sweating helps this process.

5 The digestive system

The process of breaking down the food we eat into smaller substances is called digestion. It is the function of the digestive system. The substances are then either absorbed by the body or removed from it (excreted).

Energy from food

Energy is stored within food by being 'trapped' in the chemical bonds within the food as it is being made. Food is consumed, digested, absorbed and stored within your body.

During this process, energy is made available when the chemical bonds in which it is trapped are broken. The energy is available for various processes including movement.

Your digestive system is made up of the **gastrointestinal tract** (alimentary canal) and other organs. The gastrointestinal tract is 6–8 metres long in an adult. It begins with the mouth and **oesophagus** (food pipe), then continues into the stomach, small intestine and large intestine and ends at the anus.

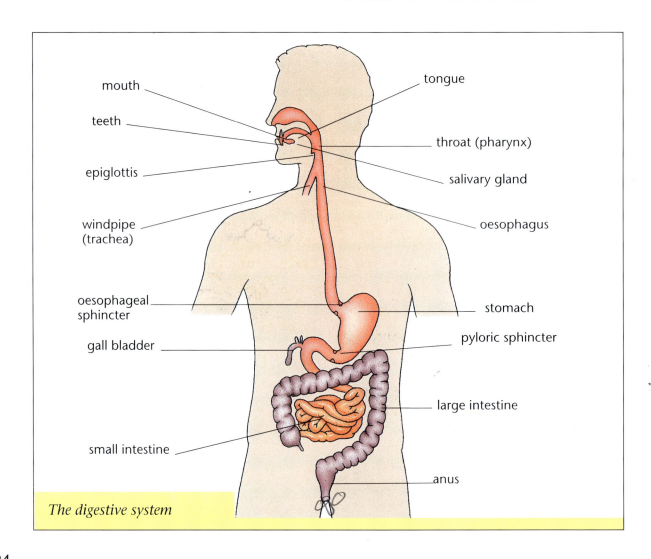

mouth
teeth
epiglottis
windpipe (trachea)
oesophageal sphincter
gall bladder
small intestine

tongue
throat (pharynx)
salivary gland
oesophagus
stomach
pyloric sphincter
large intestine
anus

The digestive system

Process of digestion

Mouth (buccal) cavity

Food is taken into your mouth (ingestion) and chewed. The teeth grind and mash the food into smaller pieces. The food is rolled into a ball which makes it easier to swallow. When food enters your mouth the **salivary glands** release extra saliva. This keeps your mouth and throat moist, moistens the food and starts to break it down. Saliva contains an enzyme, **amylase** (ptyalin). This enzyme starts the process of breaking starches (or carbohydrates) down into sugars.

During swallowing, your tongue moves up and back, pushing the food into the **pharynx** (throat). The **larynx** (voice box) is pulled up to make the **epiglottis** close over it, which prevents food from entering the lungs.

Oesophagus (gullet)

After you swallow some food, it passes through the pharynx and enters the oesophagus. The food is moved along the oesophagus by the muscular action known as **peristalsis**. This is a wave-like action of rhythmic contractions and relaxations of the muscle along the gastrointestinal tract.

Stomach

The food goes into your stomach through a one-way valve called the **oesophageal sphincter**. Your stomach acts as a temporary store for the partially-digested food. The stomach lining holds millions of **glands** that produce hydrochloric acid (HCl) and powerful, enzyme-containing digestive juices (gastric juices). These liquefy the food and break it down. Not very much food is absorbed in the stomach, but water, alcohol and aspirin are readily absorbed into the blood stream. This is why they have a rapid effect on the body.

On average, the stomach's volume is about 1.5 litres, but it can hold between 50 ml when 'empty' and 6 litres when fully distended after a particularly large meal.

The contents of the stomach are mixed with acid and digestive juices to produce a slushy acidic mixture called **chyme**. The chyme gradually empties into the small intestine. After eating, the stomach takes two to three hours to empty, depending on the size and type of meal.

Stomach (gastric) emptying is controlled by the opening and closing of another valve leading out of your stomach, called the **pyloric sphincter**. If after a large meal the stomach gets too stretched, signals are sent to the pyloric sphincter causing it to relax so that more chyme can leave the stomach and enter the small intestine. If the intestine gets too stretched the amount of chyme leaving the stomach is reduced.

Fact File

Under normal conditions saliva is released continuously from the salivary glands. However, the amount of saliva released varies depending on the conditions. When you are dehydrated, in order to save water, saliva is not released and your mouth feels dry. This makes you feel thirsty. Food in the mouth stimulates the release of saliva as does the sight, smell, sound and feel of food.

Small intestine

When it leaves your stomach the chyme enters a passage called the small intestine. This is divided into three sections: the **duodenum** (first 0.3 m), **jejunum** (next 1–2 m) and the **ileum** (last 1.5 m). Your pancreas is linked to the small intestine and releases juices to neutralize the acid in the chyme. Special enzymes break down more complex pieces of food. Fat is difficult to break down, and needs bile for proper digestion. Bile is produced in your liver and stored in your gall bladder. Your gall bladder releases bile into the duodenum when it is needed.

The broken-down food is absorbed into the blood stream, along with water, alcohol, vitamins and minerals. The food passes through the walls of the small intestine and into your blood stream. The walls of the small intestine are lined with millions of little projections called **villi**. These greatly increase the surface area between the small intestine and the blood vessels, so that more nutrients can pass into your blood and be carried to the liver. Some nutrients can be stored by the liver, and released as and when they are needed.

As the chyme moves through your small intestine (this takes from three to ten hours) propulsive movements continue to churn and mix it until it reaches the sphincter into the large intestine. The indigestible waste products (the remaining ten per cent of food) move into the large intestine.

Large intestine

This is the final section of your gastrointestinal tract. It is about 1.2 metres long and contains no villi. It is also known as the colon or bowel. Here water and minerals are re-absorbed.

Waste fluids are taken by the blood to your kidneys, where they are filtered and pass as urine through the bladder to be excreted. Undigested food residue or faeces are stored ready for excretion through the anus.

Nutrition and exercise

You must consume food, to supply your body with the nutrients it needs to survive. The six essential nutrients are:

- *carbohydrates* – which provide the body with energy;

- *fats* – which provide stored energy and insulation;

- *proteins* – which supply material for growth and repair;

- *minerals* – which contribute towards growth and repair and essential body chemistry;

- *vitamins* – to regulate body metabolism and development;

- *water* – to transport nutrients, waste products and hormones. Water is also a major component of cells and plays an important role in temperature regulation.

Foods contain various proportions of these nutrients. Bread, rice, pasta and vegetables are valuable sources of carbohydrates. Butter, margarine, oils, and cheese are high in fat. Dairy products, meat, fish and nuts are valuable sources of protein.

A healthy diet includes a variety of different foods that allow the body to function at its best. Fat is an important nutrient, but unfortunately we tend to eat too much of it. This can lead to obesity and other health problems, including heart disease and cancer.

Carbohydrate is the most important fuel for exercise, but we can only store a limited amount of it in muscles. Therefore, it is important that we eat enough carbohydrate every day to keep our stores high.

Fluids

When you exercise, your body produces a lot of heat. To prevent an excessive increase in body temperature, you must get rid of this heat. Sweating helps us to do this.

Sweating is a very effective way of losing heat but it may cause dehydration. It is possible to lose up to two kilograms of body fluid or approximately four per cent of body mass during a vigorous workout. This will have a negative effect on your performance. It is important, therefore, to take in extra fluid before, during and after exercise. However, drinking large quantities of fluid can cause discomfort so it is better to drink a little quite often.

Carbohydrate is an important fuel for exercise

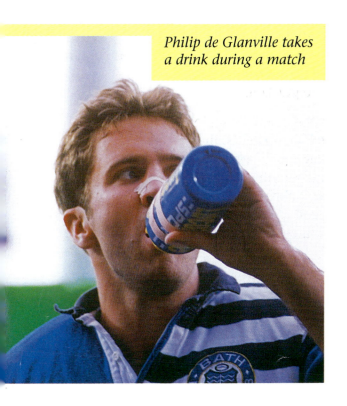

Philip de Glanville takes a drink during a match

Fact File

Pasta parties are often held for athletes the night before a marathon. This is to help ensure that they have plenty of carbohydrate in their body before the race. However, the athletes need to have been eating a high carbohydrate diet throughout their training to ensure their carbohydrate stores in the body are at their maximum.

6 The nervous system

The nervous system is one of your body's main control centres. Its function is to sense changes inside or outside the body, interpret the changes and then bring about a response. It controls all your body's movements, and has many other functions.

The nervous system is organized into:

- the **central nervous system** (CNS)
- the **peripheral nervous system** (PNS)

Central nervous system

This is the control centre for the entire nervous system. It consists of your brain and **spinal cord**. Information from around the body is sent to the central nervous system to be interpreted and acted upon.

Brain

The brain is the control centre for all voluntary (conscious) and involuntary (automatic or unconscious) activities. Most movements that you make are voluntary because you choose to move. Involuntary activities include your heart beat, which occurs without you thinking about it. The brain is a very delicate organ which needs to be protected. This protection is given by your **cranium** (skull), three layers of membrane around the brain and your **cerebrospinal fluid**. This fluid circulates round your brain, acts as a shock absorber and helps protect the brain from injury.

Your brain has many different parts each with a special function. Three important areas are:

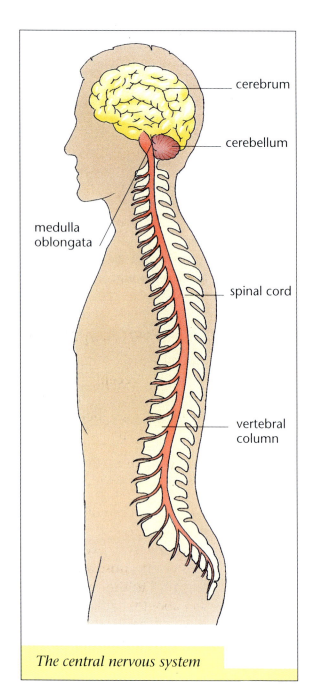

The central nervous system

- **cerebrum** – the largest portion of the brain. This is responsible for conscious control of movement as well as thought, speech, memory, imagination, learning and decision-making;

- **cerebellum** – the second largest portion of the brain. This is concerned with co-ordination of movement and controls posture and balance;

- **medulla oblongata** – the lowest part of the brain stem. This connects with the spinal cord. It is responsible for involuntary activities such as breathing, digestion and heart beat.

Spinal cord

The spinal cord is located inside your spine (vertebral column). It extends from the base of your brain to your lower back. In an adult it is about 42–45 cm long and its circumference is about 2.54 cm in the middle of the back, although it is bigger in some parts.

Peripheral nervous system

This is the link between the central nervous system and the **receptors**, muscles and **glands** around the body. It consists of a large network of nerves that spread out around the body.

The peripheral nervous system can be divided into:

- the **sensory (afferent) system** – conveys information from around the body to the central nervous system. Your ears, eyes, mouth, nose and skin all send information to the central nervous system about what you hear, see, taste, smell and feel, and it responds appropriately. These organs are called the sensory organs. During exercise you get a lot of information about your surroundings from what you see, hear and feel;

- the **motor (efferent) system** – conveys information from the central nervous system to your muscles and glands to bring about an action.

The nervous system helps these swimmers balance at the start of a race

Structure of a neurone

Nerves are made up of bundles of nerve cells or **neurones** that are surrounded by connective tissue. Neurones are the basic unit of the nervous system and are made up of three parts:

- **cell body** (**soma**) – the main cell body that contains the nucleus;

- **dendrites** –short projections from the cell body; their function is to conduct electrical signals called **nerve impulses** (information) toward the cell body;

- **axon** – a single long process that conducts nerve impulses away from the cell body. At the end, the axon branches into many fine filaments called axon terminals. Axons vary in length from a few millimetres (in the brain) to one metre (between the spinal cord and the toes).

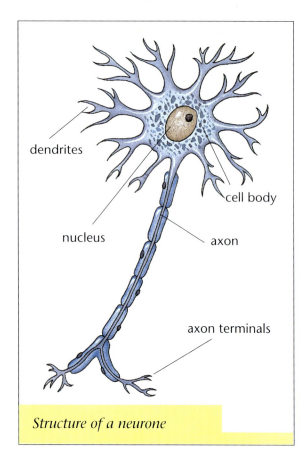

dendrites

cell body

nucleus

axon

axon terminals

Structure of a neurone

Impulse transmission

Neurones are able to generate and conduct nerve impulses. Information is conveyed along the nerve network by impulses being transmitted from one neurone to another or from a neurone to a muscle or gland. The speed at which impulses are transmitted is your speed of reaction.

Types of neurones

Some neurones are under voluntary control and some are under involuntary control:

- **voluntary control** – These neurones convey information from the central nervous system to skeletal muscle. You are able to control skeletal muscle so you can choose when to move;

- **involuntary control** – These neurones convey information from the central nervous system to smooth muscle tissue, cardiac muscle tissue and glands. This occurs without you having to think about it. The heart, blood vessels and respiratory muscles are examples of areas that are under involuntary control. This is a good thing because if you had to think about every heart beat and every breath, you would be kept very busy!

Fact File

When a sports person receives a blow to the head there is always a risk of concussion (temporary loss of consciousness). Any person experiencing concussion should be taken to a doctor for a proper examination.

Control of movement

When you want to perform a movement such as bending an arm, nerve impulses are sent from the central nervous system to your arm muscles. The impulse causes the muscles to contract and the arm to bend. Impulses are also sent to other muscles to cause them to relax so that the movement can occur. The contraction and relaxation of muscles has to be co-ordinated so that the movement is smooth and controlled. This is the job of the nervous system.

Many movements that you make are in response to information that you have received from outside the body. This information comes from the sensory organs. They allow you to see, hear and feel what is going on around you. In sport you can see where other players are, hear the referee or the crowd and feel the ball or racket. You then use your brain to decide what action to take and how to move. Impulses will then be sent to the muscles to allow this to occur.

Not all information comes from outside the body. You also get information from receptors within your body that affect how you move. There are various receptors in the muscles, joints, **tendons** and **ligaments** that sense what is going on during movement and send information to the central nervous system about the changes occurring. These receptors are called **proprioceptors**. They sense information about the amount of contraction in muscles and the position of joints in space. The information from these receptors means, for example, that you can touch the end of your nose when your eyes are shut. This is because you know where your nose and hand are without looking.

Archery requires very fine control of movement

Fact File

A reflex action is a quick, automatic response to something. An impulse is sent rapidly from a receptor to the brain or spinal cord and directly to the muscles involved. Often the brain is by-passed. Some examples of reflexes are the knee jerk reflex, removing a hand from a hot object and blinking in response to something touching the outside of the eye. These are essential reactions that protect the body from harm but not all reflexes are in response to danger.

The skeletal system or skeleton is the framework of bones and joints inside your body. The functions of the skeletal system are to provide support, shape, protection and movement, and to produce blood cells.

Support and shape

The skeleton provides support to keep your body upright as well as supporting its other components such as organs and soft tissues. It also provides a point of attachment for many muscles and gives your body its basic shape.

Protection

Parts of your skeleton surround your vital organs and protect them from damage. The skull protects the brain, the rib cage protects the heart and lungs, the spine protects the spinal cord and the pelvis protects the internal reproductive organs.

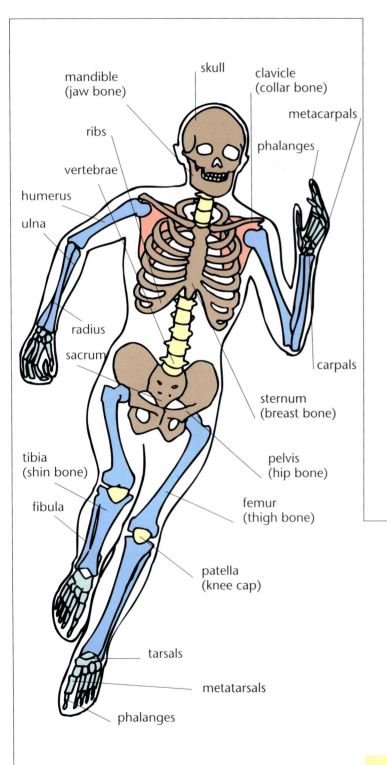

mandible (jaw bone)
skull
clavicle (collar bone)
metacarpals
phalanges
ribs
vertebrae
humerus
ulna
radius
sacrum
carpals
sternum (breast bone)
pelvis (hip bone)
tibia (shin bone)
fibula
femur (thigh bone)
patella (knee cap)
tarsals
metatarsals
phalanges

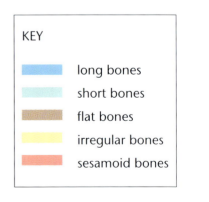

KEY

long bones
short bones
flat bones
irregular bones
sesamoid bones

The skeletal system

Movement

This takes place as a result of muscles contracting and pulling a bone about a joint. The amount of movement that can occur at a joint depends on its structure. Some joints are designed for a large range of movement, others for little or no movement.

Blood cell production

Red blood cells are produced in the middle part, the marrow, of certain bones. Some white blood cells and platelets are also made in the marrow. The bone also stores important mineral salts such as calcium and phosphorous.

Divisions of the skeleton

The adult skeleton is made up of 206 bones that can be divided into two parts:

- **axial** skeleton consisting of the skull, the thoracic cage and the vertebral column (spine);

- **appendicular** skeleton consisting of the bones of the arms and legs and of the shoulder and hip girdles that link them to the axial skeleton.

Classification of bones

Within the body there are four principal classes of bones:

- long bones
- short bones
- flat bones
- irregular bones

Long bones

These bones are longer than they are wide and are slightly curved for strength. The curve allows the bone to absorb the stress of the body weight at certain points and this helps prevent breaks. The femur (thigh), fibula and tibia (lower leg), humerus (upper arm), ulna and radius (forearm) and phalanges (fingers and toes) are all long bones. These bones act as levers and play an important role in movement.

Short bones

These are almost cube-shaped because their length and width are nearly equal. The carpals (in the wrist) and tarsals (in the ankle) are short bones.

Flat bones

These are thin, flat bones that are very important for protection. They also provide a large surface area where muscles are attached. The pectoral (shoulder) girdle, pelvic (hip) girdle, the ribs and the cranium (skull) are all flat bones.

Irregular bones

These are irregular-shaped bones that cannot be grouped into one of the above categories. The vertebrae and certain facial bones are classed as irregular bones.

Sesamoid bones are classified because of where they are rather than because of their shape. These bones are located in tendons where considerable pressure develops. The patella (knee cap) is an example of this type of bone.

Joints

A joint is where two bones meet, where **cartilage** and bone meet or where teeth and bone meet. Movement is produced when a muscle contracts and moves a bone (lever) about a central point (fulcrum). Joints are the central point around which bones move. Joints can be classified according to how much movement they allow:

- **immovable joints** are fixed joints at which no movement occurs. The joints between the bones in the skull are examples of immovable joints. Such joints exist in young people to allow growth but then become solid in adults;

- **slightly movable joints** permit only a little movement. Examples are the joints between vertebrae and the joint between the tibia and fibula at the ankle end of the lower leg;

- **freely movable joints** are **synovial joints** and permit free movement. Due to the movement permitted, these joints play an important role in any form of physical activity or exercise.

We shall now examine some of the tissues that help form the structure of joints in general.

Connective tissues

The main function of the connective tissues is to bind and support various structures in the body. There are many different types of connective tissues. Three types that relate to joints are:

- **cartilage** which consists of a dense network of fibres that are capable of absorbing a lot of force. There are different types of cartilage present in the body.

Hyaline cartilage is an important component of joints. It covers the joining surfaces of the bones in the joints and its job is to protect the bone and reduce friction in the joint. Other types of cartilage help maintain the shape of certain organs such as the nose, the flaps of the ear and the epiglottis of the larynx;

- **ligaments** which are bands or cords of fibres that join bone to bone across a joint and restrict its movement. As a joint nears the end of its range of motion, the ligaments pull tight and stop further movement. When a joint is forced beyond its normal range of movement, the surrounding ligaments may be strained or torn. This makes the joint less stable;

- **tendons** which are tough bands or cords of densely packed fibres similar to ligaments but more elastic. They always make up part of a muscle and join muscle to bone allowing movement to take place.

Synovial joints

These joints have a space between the bones called the synovial cavity. This allows a greater range of movement than in other joints. There are six different types of synovial joint but they have certain common characteristics:

- **articular capsule** – this is a sleeve-like covering that encloses the synovial cavity and connecting bones. It is composed of two layers, an outer layer of fibres and an inner layer called the synovial membrane.

The outer layer consists of dense connective tissue that is flexible enough to allow movement yet strong enough to resist dislocation.

The synovial membrane releases synovial fluid into the joint and this acts as a lubricating fluid. This helps reduce friction and allows the joint to move more easily. The fluid also provides nutrients to the hyaline cartilage.

- hyaline cartilage – this is the type of cartilage that covers the surfaces of the connecting bones;

- **ligaments** – these join bone to bone and limit the range of movement in the joint;

- **articular discs (menisci)** – these are pads of fibro-cartilage that lie between the connecting surfaces and are attached to the fibrous capsule. They allow bones of different shapes to fit neatly together and also act as shock absorbers.

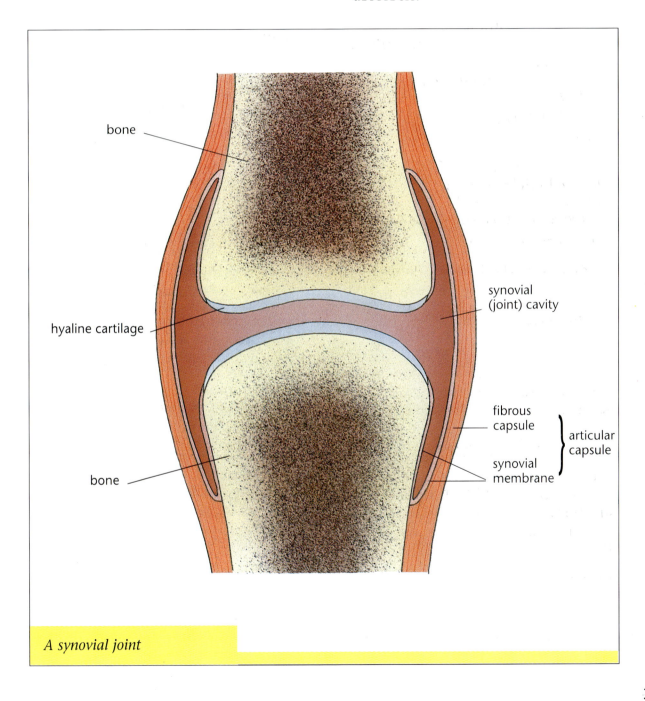

A synovial joint

Types of synovial joints

There are six types of synovial joints, each allowing different types and ranges of movement:

- *gliding joints* – joining surfaces are usually flat and glide over each other. They allow side-to-side and back-and-forth movements. They include the joints between the tarsals and the carpals, the sternum and clavicle and the scapula and clavicle;

- *hinge joints* – a convex surface of one bone fits into the concave surface of another. They only allow movement in one direction, for example bending and straightening the elbow and finger joints;

- *pivot joints* – a rounded, pointed or cone-shaped surface of one bone fits into a ring formed by another bone and a ligament. They allow rotation, for example the joint near the elbow between the two bones in the forearm that allows the forearm to rotate;

- *condyloid or ellipsoidal joints* – an oval-shaped condyle of one bone fits into an elliptical cavity of another. They allow side-to-side and back-and-forth movements. For example, the joint between the radius and carpals in the wrist,

- *saddle joints* – a special form of condyloid joint where the joining bones are shaped like a rider and a saddle. They allow side-to-side and back-and-forth movements. For example, the joint between the bones at the base of the thumb;

- *ball and socket joints* – a ball-like bone fits into a socket-like surface of another bone. They allow movement in three directions: side-to-side, back-and-forth and rotation. For example, the hip and shoulder joints.

Range of movement and stability

The range of movement at a joint and its stability are related to each other. The immovable or slightly movable joints permit little or no movement but are much more stable than the freely movable joints. The ball and socket joints allow the greatest range of movement but are the least stable joints. The range of movement at the hip is less than at the shoulder, because of the joint surfaces and strong ligaments. The shoulder joint, in particular, is fairly unstable and so dislocation of the shoulder is a fairly common injury. This is especially the case in activities such as canoeing and contact sports, where strong movements occur at extreme positions.

The range of movement at a joint is determined by three factors:

- *muscles and tendons* – tightness of muscles limits range of motion, but this can be improved through flexibility training;

- *the shape of the joining surfaces* – close fitting bony structures limit the range of motion of a joint;

- *restraining effects of ligaments* – these limit range of motion and help maintain stability.

The flexibility of a joint should not exceed the ability of the muscles to support and maintain the stability of the joint.

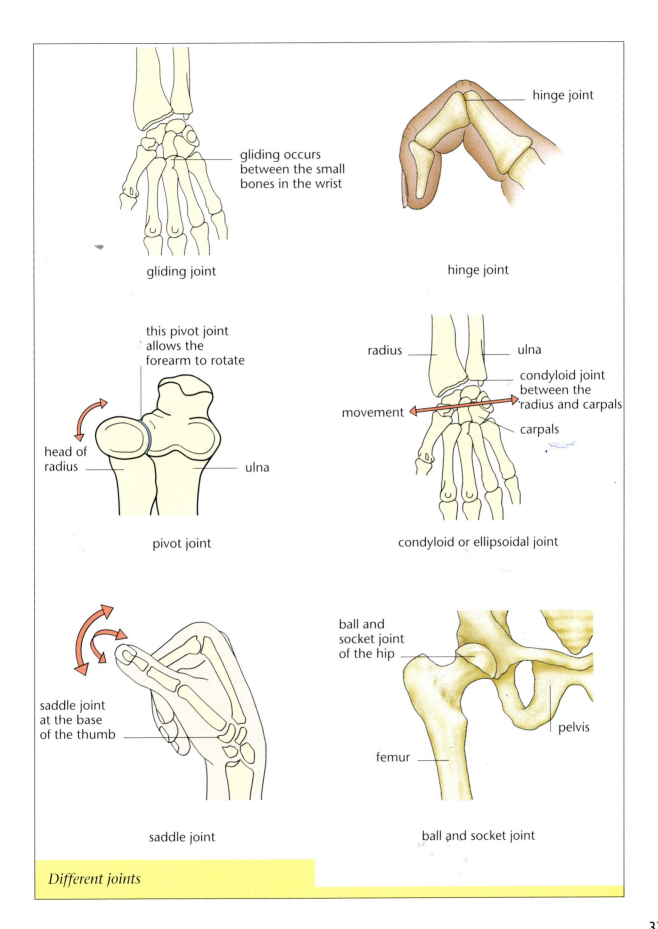

gliding occurs
between the small
bones in the wrist

gliding joint

hinge joint

hinge joint

this pivot joint
allows the
forearm to rotate

head of
radius

ulna

pivot joint

radius

ulna

condyloid joint
between the
radius and carpals

movement

carpals

condyloid or ellipsoidal joint

saddle joint
at the base
of the thumb

saddle joint

ball and
socket joint
of the hip

pelvis

femur

ball and socket joint

Different joints

The vertebral column

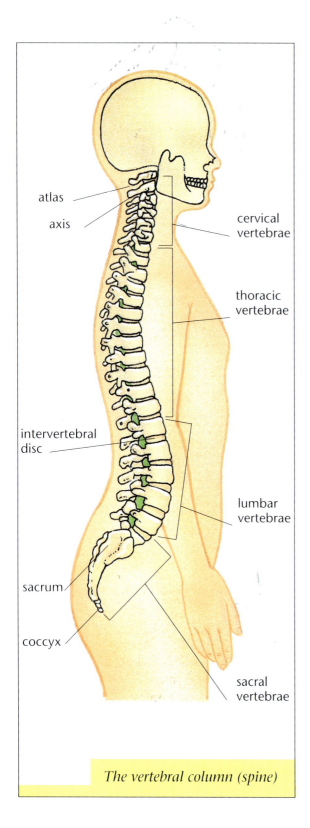

atlas

axis

intervertebral disc

sacrum

coccyx

cervical vertebrae

thoracic vertebrae

lumbar vertebrae

sacral vertebrae

The vertebral column (spine)

Your vertebral column or spine is composed of a series of vertebrae and makes up about two-fifths of the total height of your body.

The functions of the vertebral column are to:

- keep the body upright

- enclose and protect the spinal cord

- support the head

- serve as a point of attachment for the ribs and muscles of the trunk

- act as a shock absorber

- contribute to movement.

The spine is divided into five regions:

- *cervical region* – there are seven cervical vertebrae located in the neck area. The first vertebra, the atlas, supports the head and the second vertebra, the axis, is the pivot on which the atlas and head rotate. The neck muscles attach here and movements of the head such as nodding take place in this area;

- *thoracic region* – this consists of twelve large, strong vertebrae in the chest area. These vertebrae link with the ribs and support the rib cage (except for the bottom two thoracic vertebrae). Some movement is possible in this area, but not much;

- *lumbar region* – this consists of five vertebrae which are the largest and strongest in the spine. They are located in the lower back area and their main job is to support the weight of the body. Lumbar vertebrae are well adapted for muscle attachment. Most of the back muscles attach in this area;

- *sacral region (sacrum)* – this consists of five fused (joined together) vertebrae that form a triangular bone. The sacrum joins with the pelvic girdle and transfers weight from the trunk to the hips and legs. Fusion of these bones begins at 16–18 years and should be complete by the mid-twenties;

- *coccyx* – this consists of four fused vertebrae and is also triangular-shaped. It is often referred to as the tail as it is located at the very bottom of the vertebral column. This region of the spine serves no function.

Types of movements

Various types of movement are possible at different joints. Many of these movements occur in pairs and oppose each other.

General movements

General movements are possible at a number of joints:

- *abduction* – movement of a bone away from the mid-line of the body. For example, lifting your leg out to the side;

- *adduction* – movement of a bone towards the mid-line of the body. For example, bringing your leg back in from the side;

- *extension* – increasing the angle between two bones. For example, straightening your elbow;

- *flexion* – decreasing the angle between two bones. For example, bending your elbow;

- *hyperextension* – the extension past the starting or straight position.

For example, many females can hyperextend their elbow past the straight-arm position;

- *inward (medial) rotation* – rotating a limb towards the mid-line of the body. For example, turning your arm inward;

- *outward (lateral) rotation* – rotating a limb away from the mid-line of the body. For example, turning your arm outward.

Special movements

Special movements only occur at a particular joint:

- **dorsiflexion** – bending the foot up towards the shin;

- **plantarflexion** – bending the foot downward away from the body in the direction of the sole;

- **eversion** – turning the sole of the foot outward;

- **inversion** – turning the sole of the foot inward;

- **pronation** – turning the forearm so that the palm faces downward or backward;

- **supination** – turning the forearm so that the palm faces upward or forward.

Most movements that you make involve a combination of the above movements. For example, kicking a ball involves knee flexion and extension, hip flexion and extension, foot plantarflexion and dorsiflexion. In addition, movements of the arms and trunk will also take place. It is the job of the nervous system to ensure all these movements occur in a co-ordinated manner.

8 The muscular system

Bones and joints provide the levers and framework for movement but it is the muscles that are able to produce the movement. They do this by contracting and relaxing. Muscle contractions also help maintain posture and produce heat to maintain body temperature. There are three types of muscle in the body:

A weight lifter's skeletal muscles are strong and welldefined

- **smooth muscle** – this type of muscle is involved in control of things inside the body. Its contraction is not under voluntary (conscious) control and so it is called involuntary. It is found in the walls of hollow structures such as the intestines and the blood vessels;

- **cardiac muscle** – this special type of muscle makes up the walls of the heart. It is not found in any other part of the body. It is striped in appearance and is involuntary, like smooth muscle. Cardiac muscle is able to contract and relax without external nerve stimulation, so that it can keep on beating by itself;

- **skeletal muscle** – this type of muscle is mainly attached to the skeleton. It is striped in appearance and is voluntary. That means it can be made to contract and relax by conscious control.

The term 'muscular system' refers to the skeletal muscles of the body. Cardiac muscle forms the heart so is part of the circulatory system. The location of

smooth muscle determines which system it is part of. Smooth muscles in the intestines are part of the digestive system whereas smooth muscles in the blood vessels are part of the circulatory system.

Skeletal muscle is made up of bundles of muscle fibres. These fibres are cylindrical cells that lie parallel to each other. Muscle fibres are surrounded and separated from each other by a layer of **connective tissue** (endomysium). Each bundle of fibres is surrounded and separated from other bundles by another layer of connective tissue (perimysium). Finally, each muscle is surrounded by a further layer of connective tissue (epimysium). These layers of connective tissue continue beyond the length of the muscle and group together to form a **tendon** that then connects the muscle to a bone.

Special terms are used to describe the points at which the tendon at either end of the muscle attaches to the bone:

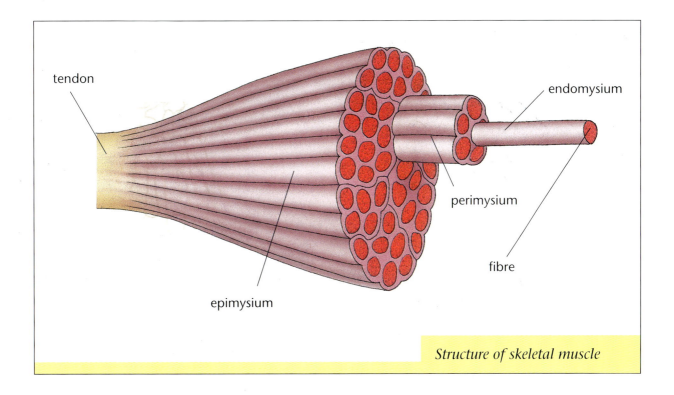

tendon

endomysium

perimysium

fibre

epimysium

Structure of skeletal muscle
Structure of skeletal muscle

- **origin** – the point of attachment on the bone at the end nearest the centre of the body;

- **insertion** – the point of attachment on the bone at the end furthest from the centre of the body.

Muscle fibre types

There are two main types of muscle fibres that are distinguished from each other by their speed of contraction:

- **slow twitch fibres** are so named because they contract relatively slowly. They cannot supply high levels of energy at once but they can supply a lot of energy over a period of time and are resistant to fatigue. These fibres are therefore used for endurance type activities;

- **fast twitch fibres** are able to contract very quickly and produce a lot of energy in an instant. However, they tire quickly. These fibres are used for activities that require speed and power, such as sprinting and jumping.

Fact File

Muscles are composed of a mixture of both muscle fibre types but the proportions vary from person to person. These proportions appear to be relatively fixed for each person. People with high percentages of slow twitch fibres will be better at endurance type activities. People with high percentages of fast twitch fibres will be better at power events such as sprints and jumps.

Control of movement

Your muscles contract when they are stimulated by the nervous system. When you decide to make a movement such as flexing a knee, a **nerve impulse** is sent from the brain to the relevant leg muscles, the hamstrings. This impulse causes the muscles to contract and shorten, which causes the knee to flex.

These dancers demonstrate good control of movement

Muscles are only able to pull a bone towards another bone, they cannot push a bone away. This is why they work in pairs creating opposite movements.

The quadriceps muscles when shortened cause the opposite movement to the hamstrings by extending the leg. It is important, therefore, that these muscles do not work against each other. The nervous system makes sure this does not happen. It does this by sending a nerve impulse to the quadriceps causing them to relax at the same time as it is sends the nerve impulse to the hamstrings to contract.

Throughout your body at the various joints, opposing muscles work in this way to allow movement to occur. The special terms used to describe the different roles of the muscles are:

- **agonist** (**prime mover**) – the muscle that contracts to produce the movement. In the above example this is the hamstrings;

- **antagonist** – the muscle that relaxes to allow the movement to occur. In the above example this is the quadriceps.

For the opposite movement, extending the leg, the muscles reverse roles. The quadriceps are now the agonist as they cause the leg to extend by contracting. The hamstrings are the antagonist as they relax to permit the movement. Muscles that work in this way are described as an **antagonistic pair**.

Muscles that assist the agonist and reduce any unnecessary movements are called **synergists**. Muscles that help stabilize the origin of the agonist are called **fixators**.

Types of muscle contractions

Normally we tend to think of a contraction as resulting in a muscle shortening. However, there are three types of muscle contraction that do not all result in the muscle shortening:

- **concentric contraction** – the muscle shortens as it develops tension and overcomes a resistance. For example in a biceps curl the biceps contracts and shortens to pull the hand towards the body;

- **isometric contraction** – in this type of contraction no movement occurs around a joint but tension is still developed in the muscle. If you hold a heavy weight with the arms bent at right angles, as for a biceps curl, and try to keep the weight in the same position, then this is an isometric contraction. The weight acts to try to straighten the arms, therefore, the biceps muscles must resist this by developing tension and keeping the weight in the same position;

- **eccentric contraction** – the muscle lengthens as it develops tension. In a biceps curl when the weight is lowered the biceps gradually lengthen. The biceps develops tension in order to control that lengthening. This is an eccentric contraction.

Muscle tone

At any given time some fibres in a muscle are contracted while others are relaxed. The number of contracting fibres may not be enough to produce movement but they do tighten the muscle. This is called muscle tone. Muscle tone develops when stretch develops in a muscle. Stretch receptors sense the amount of stretch and

You can see the tension in this gymnast's arms as he holds the crucifix position

respond by developing tone in the muscle. This helps to maintain position. For example, the muscles in the back of the neck are partially contracted (toned) to prevent the head from falling forward and to keep it in the correct position. When you are asleep, all fibres in the muscle relax and there is no muscle tone. When you are stressed, muscle tone can increase and cause stiffness in such areas as the neck and shoulders.

Posture

Posture is the way you hold your body. Your body needs to be held in a position that allows its different parts to function properly. Muscle tone is very important in maintaining posture.

Good posture is important for:

- sitting, standing and moving around
- activities such as dance and gymnastics where the performer wants to look good
- health
- the prevention of injuries.

Poor posture can result in damage to surrounding organs and tissues. Many people have lower back problems that are often due to poor posture. Slouching instead of keeping the back straight, and incorrect posture when lifting objects can result in postural problems and injuries.

Stretching and strengthening the muscles that control posture can help prevent these problems.

Categories of muscles

Just as the movements at joints were categorized in the last chapter, the muscles that produce those movements can be categorized in a similar fashion:

- *abductors* – the muscles that move a limb away from the mid-line of the body;
- *adductors* – the muscles that move a limb back towards the mid-line of the body;
- *extensors* – the muscles that increase the angle between two bones;
- *flexors* – the muscles that decrease the angle between two bones;
- *rotators* – the muscles that cause a limb to rotate.

Muscle	Location	Function
Triceps	At the back of the upper arm	Elbow extension
Biceps	Front of the upper arm	Elbow flexion
Deltoids	On the shoulders	Shoulder movements
Pectorals	Front of the upper chest	Assist shoulder and upper arm movements
Trapezius	At either side of the neck on the upper back	Shoulder blade movements and head extension
Gluteals	The bottom	Abduction, adduction, rotation and extension of thigh
Quadriceps	Front of the upper leg	Knee extension and hip flexion
Hamstrings	Back of the upper leg	Knee flexion and hip extension
Gastrocnemius	Back of the lower leg	Plantarflexion of the foot and knee flexion
Latissimus dorsi	On the back from the middle of the rib cage to the lower back	Upper arm movements. Draws arm downwards and backwards
Abdominals	Front and side of the stomach	Flexion and rotation of the trunk and to assist breathing and posture

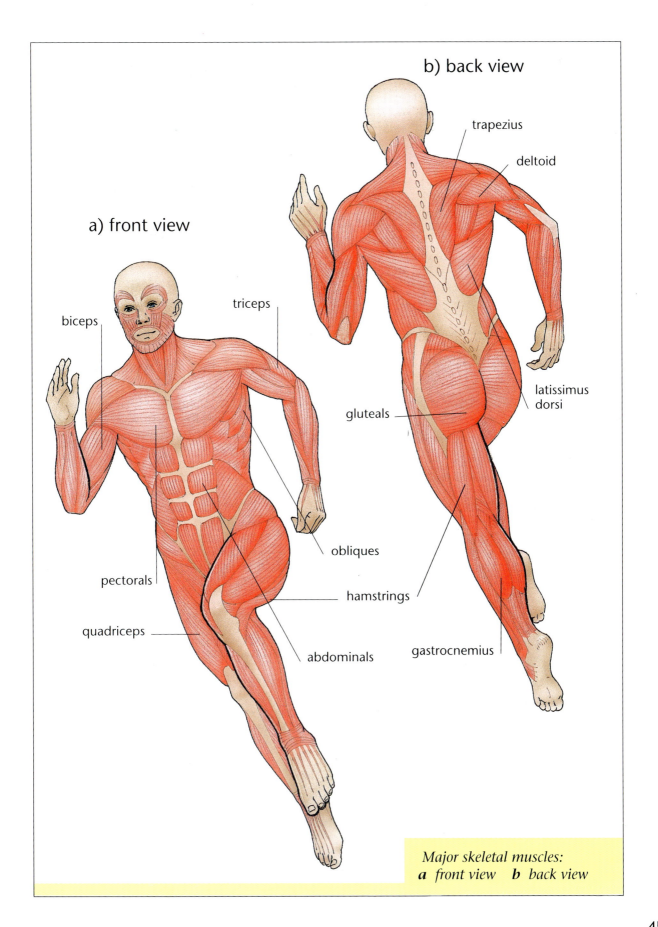

a) front view

biceps

triceps

pectorals

quadriceps

obliques

hamstrings

abdominals

b) back view

trapezius

deltoid

latissimus
dorsi

gluteals

gastrocnemius

Major skeletal muscles:
a *front view* **b** *back view*

Glossary

adenosine triphosphate (ATP) usable form of energy for exercise

aerobic energy energy obtained with the use of oxygen

agonist (prime mover) the name given to a muscle that contracts to produce a movement

alveoli tiny air sacs that form part of the lungs

amylase an enzyme in saliva that starts the breakdown of starches into sugars

anaerobic energy energy obtained without the use of oxygen

antagonist the name given to a muscle that relaxes to allow a movement to occur

aorta large blood vessel that carries oxygenated blood out of the left ventricle of the heart on its way around the body

arteries blood vessels that carry blood away from the heart

arterioles small arteries that branch off from arteries and lead to the tissues of the body

articular capsule a sleeve-like covering that encloses synovial joints, composed of a fibrous layer and a synovial membrane

articular discs (menisci) pads of fibro-cartilage that lie between the connecting surfaces in a joint

atherosclerosis accumulation of fat on the walls of an artery

atria (or auricles) the two upper chambers of the heart, right atrium and left atrium

bicuspid (mitral) valve a one-way valve in the heart separating the left atrium from the left ventricle

blood pressure the force that blood exerts on the walls of the blood vessels

bronchi plural of bronchus

bronchioles small passageways that branch off from the bronchi

bronchus passageway for air from the trachea (windpipe) to a lung

buccal cavity mouth cavity

capillaries tiny blood vessels that link arterioles and venules and allow exchange of materials between blood and body cells

cartilage a dense network of fibres that are capable of absorbing considerable stresses

cell the basic unit of life that can maintain itself, grow and reproduce

cerebellum second largest portion of the brain, responsible for co-ordination of movement

cerebrospinal fluid circulates around the brain and spinal cord to act as a shock absorber

cerebrum the largest portion of the brain, responsible for conscious control of the body

chyme a slushy acid mix produced in the stomach from food, acid and digestive juices

concentric contraction a type of muscle contraction during which the muscle shortens as it develops tension

connective tissue type of tissue that binds parts of the body together

cranium the skull, encloses and protects the brain

diaphragm large dome shaped muscle beneath the lungs that plays an important role in breathing

diastolic blood pressure pressure of the blood when the heart is relaxing

diffusion movement of particles from an area of high concentration to an area of low concentration

duodenum the first section of the small intestine

eccentric contraction a type of muscle contraction during which the muscle lengthens as it develops tension

epiglottis a flap at the top of the larynx that allows air in and keeps food out

erythrocytes red blood cells

expiratory reserve volume the amount of air that can be forcefully expired at the end of a normal expiration

fixator the name given to a muscle that helps stabilize the origin of the agonist (prime mover)

gastrointestinal tract (alimentary canal) a tube running from the mouth to the anus

gland an organ or cell that makes chemical substances and releases them for the body to use or excrete

haemoglobin a component of red blood cells

haemophilia a bleeding disease in which clotting does not occur, can result in major blood loss

hyaline cartilage the type of cartilage that covers the surface of connecting bones

ileum the final section of the small intestine

insertion the point of attachment of a muscle on a bone at the end furthest from the centre of the body

inspiratory reserve volume the amount of air that can be forcefully inspired at the end of a normal inspiration

internal and external intercostals muscles between the ribs that are involved in breathing

isometric contraction a type of muscle contraction during which no movement occurs around a joint but tension is still developed in the muscle

jejunum the middle section of the small intestine

lactic acid a by-product formed when glycogen is broken down without the use of oxygen (anaerobic) larynx voice box, connects pharynx (throat) to the trachea (windpipe)

leucocytes white blood cells

ligament a band or cord of fibrous tissue that joins bone to bone across a joint and limits the range of movement

lumen the centre of an artery or vein

medulla oblongata part of the brain that connects with the spinal cord, responsible for involuntary activities such as breathing

mucous membrane thin layer of tissue that releases mucus

mucus a thick fluid

nasal cavity nose

nerve impulse an electrical signal that sends information along a nerve

neurone a nerve cell, the basic unit of the nervous system

oesophageal sphincter a one-way valve between the oesophagus and the stomach

oesophagus passageway for food connecting the throat with the stomach

organ a collection of different tissues that acts as a functional unit

organ system a group of organs that join together to perform a certain function

origin the point of attachment of a muscle on a bone at the end nearest the centre of the body

oxygen debt the extra oxygen taken in during recovery from exercise

oxygen uptake (VO^2) the amount of oxygen that the body takes in and uses

oxyhaemoglobin formed when oxygen combines with haemoglobin to be transported round the body; gives blood its red colour

palate at the top of the mouth, separates the nose from the mouth

peak expiratory flow the maximum flow rate at which air can be expired from the lungs

peristalsis the way in which muscular actions produce movement of food along the oesophagus

pharynx throat

plasma component of blood that contains water, proteins and other substances

pleural cavity the space between the layers of the pleural membrane

pleural membrane a double layered membrane that surrounds the lungs

proprioceptors receptors in muscles, joint, ligaments and tendons that sense information regarding the degree of contraction of muscles and the position of joint in space

pulmonary artery large blood vessel that carries deoxygenated blood out of the right ventricle of the heart to the lungs

pulmonary vein large blood vessel that carries oxygenated blood into the left atrium of the heart from the lungs

pyloric sphincter a one-way valve between the stomach and the small intestine

receptors parts of the nervous system that are sensitive to changes

residual volume the volume of air that is left in the lungs after a maximum expiration

salivary glands glands that release saliva into the mouth to moisten the mouth and throat and to moisten food and start to break it down

semilunar valve a one-way valve in the aorta and pulmonary artery that stops blood flowing back into the heart

sphygmomanometer a device for measuring blood pressure

spinal cord nerve tissue located in the spinal column

spirometer a device for measuring gas volumes

superior and inferior vena cava large blood vessels that carry deoxygenated blood from around the body into the right atrium of the heart

synergist a muscle that assists the agonist (prime mover) by reducing any unnecessary movements

synovial joint joint with a space (synovial cavity) in between the bones that allows a large range of movement

systolic blood pressure pressure of the blood when the heart is contracting

tendons tough bands or cords of densely packed fibres that are part of a muscle and join muscle to bone

thoracic cage the rib cage, protects the heart and lungs

thrombocytes component of blood also called platelets

tidal volume the amount of air that is breathed in or out in one breath

tissue a collection of similar types of cells

trachea the windpipe, a passageway for air connecting the throat to the lungs

tricuspid valve a one-way valve in the heart separating the right atrium and right ventricle

vasoconstriction decrease in the size of the lumen of a blood vessel

vasodilation increase in the size of the lumen in a blood vessel

vein blood vessel that carries blood to the heart

ventricles the two lower chambers of the heart, right and left ventricle

venules small veins that unite to connect capillaries to veins

villi special protruding structures that line the small intestine and through which compounds are absorbed into the blood stream

vital capacity the maximum amount of air that can be expired after a maximum inspiration

Index